SCIENTIFIC AMERICAN | EDUCATIONAL PUBLISHING

SCIENTIFIC AMERICAN INVESTIGATES
FOOD CHAINS AND FOOD WEBS

FOOD CHAINS IN YOUR BACKYARD

ROSIE BANKS

Published in 2026 by The Rosen Publishing Group
in association with Scientific American Educational Publishing
2544 Clinton Street, Buffalo NY 14224

Names: Banks, Rosie, 1978-.
Title: Food chains in your backyard / Rosie Banks.
Description: Buffalo, New York : Scientific American Educational Publishing, an imprint of Rosen Publishing, 2026. | Series: Scientific American investigates food chains and food webs | Includes glossary and index.
Identifiers: ISBN 9781538312490 (pbk.) | ISBN 9781538312506 (library bound) | ISBN 9781538312513 (ebook)
Subjects: LCSH: Food chains (Ecology)–Juvenile literature. | Urban Ecology (Biology)–Juvenile literature.
Classification: LCC QH541.15.F66 B36 2026 | DDC 577.5'6–dc23

Portions of this work were originally authored by Katie Kawa and published as *Backyard Food Chains*. All new material in this edition is authored by Rosie Banks.

Cover Designer: Nicholas Switalski
Interior Designer: Rachel Rising
Editor: Therese Shea

Photo credits: Cover, p. 1 Jayne Lipkovich/Shutterstock.com; p. 5 (bird) Danita Delimont/Shutterstock.com; p. 5 (food chain) AZ Septian/Shutterstock.com; p. 6 Inga Nielsen/Shutterstock.com; p. 7 VesnaArt/Shutterstock.com; p. 8 Kitthanes/Shutterstock.com; p. 9 Daniel Sobolic/Shutterstock.com; p. 11 Sergey V Kalyakin/Shutterstock.com; p. 13 LedyX/Shutterstock.com; p. 14 Bonnie Taylor Barry/Shutterstock.com; p. 15 Wirestock Creators/Shutterstock.com; p. 17 (raccoon) Harry Collins Photography/Shutterstock.com; p. 17 (crows) StoneMonkeyswk/Shutterstock.com; p. 18 kungfu01/Shutterstock.com; p. 19 Rico Aquin/Shutterstock.com; p. 20 Potapov Alexander/Shutterstock.com; p. 21 BlueRingMedia/Shutterstock.com.

Printed in the United States of America

CPSIA compliance information: Batch #CSSA26. For Further Information contact Rosen Publishing at 1-800-237-9932.

CONTENTS

Words in the glossary appear in **bold** type the first time they are used in the text.

IN YOUR OWN BACKYARD

When we think of **habitats**, we often think of places far away. Rainforests, deserts, and mountaintops are habitats for plants and animals. But so are places much, much closer. Your community parks are full of organisms, or forms of life. Even your backyard can be a habitat!

True habitats are places where many plants and animals make homes, **reproduce**, and find food. That means every habitat, even your yard, is a place to see a food chain or food web in action.

FUN FACT

FOOD CHAINS ARE A WAY TO SHOW HOW CERTAIN PLANTS AND ANIMALS PASS ENERGY TO ONE ANOTHER. A GROUP OF CONNECTED FOOD CHAINS IS CALLED A FOOD WEB.

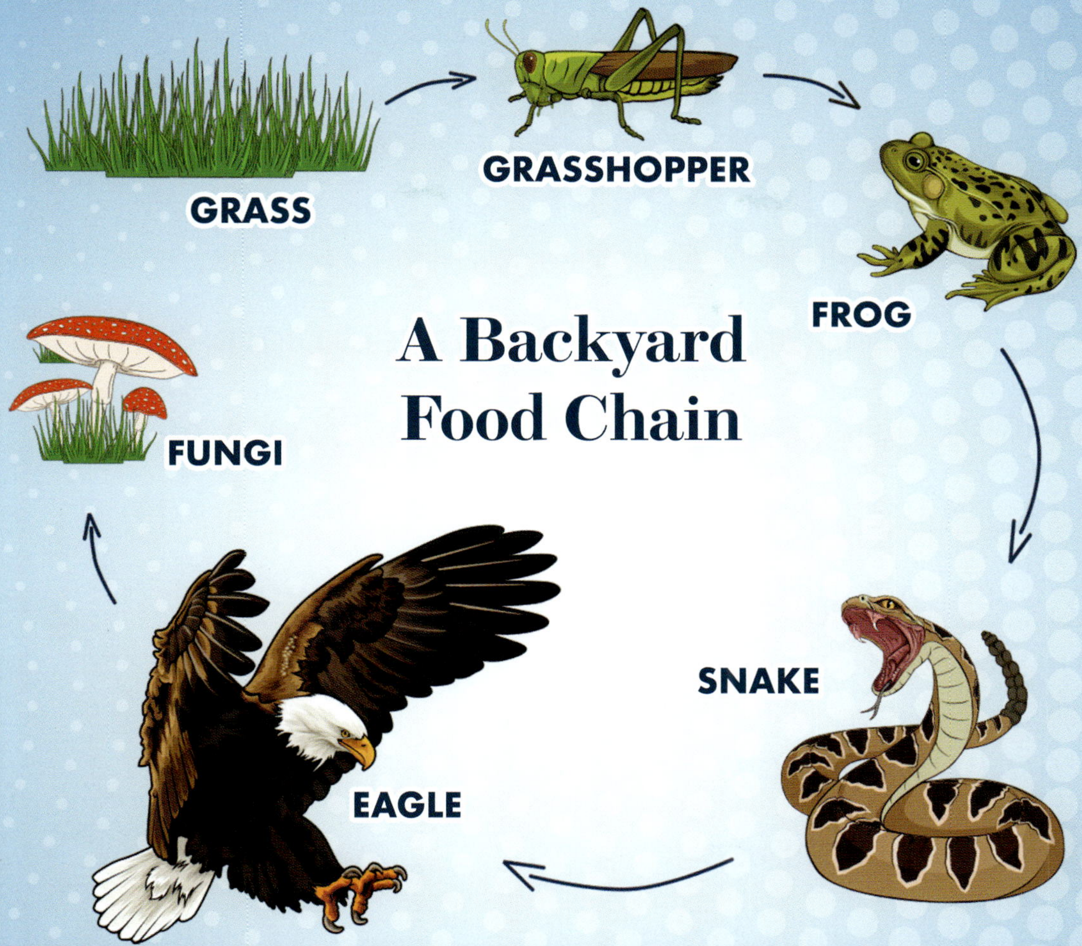

A Backyard Food Chain

If you see animals such as birds in your backyard, you might have a backyard habitat!

NATURE'S CLASSROOM

Backyard habitats provide a perfect place to learn about food chains and food webs. People can watch from inside their own homes as energy passes from a berry bush to a bird as the bird eats the berries.

People leave their mark on backyard habitats in many ways. They plant things that animals use for food such as flowers and grasses. Some people grow food for themselves in their backyard too. They may have fruit trees and vegetable gardens.

When people plant a vegetable garden, they make themselves a part of a backyard food chain!

FUN FACT

PEOPLE INTRODUCE ANIMALS INTO BACKYARD HABITATS WHEN THEY GET PETS, SUCH AS DOGS AND CATS.

HERE COMES THE SUN

Solar energy, or sunlight, is the starting point of all food chains and food webs. The sun is where the energy first comes from. Without the sun, food chains and food webs—and any form of life—wouldn't be possible on Earth.

Backyard plants get their energy from the sun. They use this energy to live and grow, but some energy is stored in the plant. Rabbits and deer eat plants that grow in backyards. When animals eat these plants, energy is passed on to the animals.

Many people find it hard to keep animals such as deer, rabbits, and groundhogs out of their gardens.

FUN FACT

MOST FOOD CHAINS SHOW ONLY ORGANISMS, SO THE SUN ISN'T INCLUDED.

PRODUCING FOOD

With the help of the sun, plants produce, or make, their own food energy. They're called producers because of this. Producers are the first link in a backyard food chain.

Plants turn sunlight, or solar energy, into energy that's part of organisms. Plants turn sunlight, water, and **carbon dioxide** into a kind of sugar. They do this through a process called photosynthesis (foh-toh-SIHN-thuh-suhs). A backyard is often filled with examples of plants using energy from the sun to live and grow. Those with vegetable or flower gardens have even more producers.

All plants in a backyard—including flowers, vegetables, grasses, trees, and bushes—use photosynthesis to make food.

FUN FACT

ALGAE AND SOME KINDS OF BACTERIA ALSO MAKE FOOD USING PHOTOSYNTHESIS. ALGAE MAY GROW IN BACKYARD PONDS.

WHO'S HUNGRY?

Many other kinds of organisms—including all animals—need to consume, or eat, their energy. They're called consumers because they consume plants and animals.

Animals that eat plants are the next link in a backyard food chain. Squirrels and caterpillars are just two of the many backyard animals that get their energy from plants. Plant-eating animals are called herbivores.

Plants need herbivores too. Some animals spread seeds through their **behaviors**. Squirrels bury seeds for winter, for example. Some of these seeds grow to become new plants.

Butterflies and flowers help each other. Flowers provide **nectar** for butterflies to drink. Butterflies carry pollen from one flower to another, which allows flowers to reproduce.

FUN FACT

ANIMALS THAT SPREAD SEEDS ARE CALLED SEED DISPERSERS. SOME SPREAD SEEDS THROUGH THEIR POOP!

WHO'S BEING HUNTED?

Backyards aren't scary places, but some animals are in danger of being eaten! Backyards are home to carnivores, or animals that eat other animals. They're the next link in the food chain. Birds are common backyard carnivores. They kill and eat bugs. Robins like to eat caterpillars, for example.

Robins are food for other carnivores—including hawks and owls. Carnivores that eat herbivores are sometimes called secondary consumers. They're yet another link in a backyard food chain.

ROBIN

Foxes are secondary consumers. They mostly eat herbivores such as rabbits and mice.

FUN FACT

ANIMALS THAT EAT SECONDARY CONSUMERS—OR OTHER CARNIVORES—MAY BE CALLED TERTIARY CONSUMERS. TERTIARY MEANS "THIRD."

THEY EAT IT ALL

Omnivores are animals that eat both animals and plants. Some will eat anything they can find, including people's trash! Raccoons are common backyard—and front yard—omnivores. They eat out of garbage cans, but they also eat nuts, berries, bugs, eggs, mice, seeds, and many other things.

Crows are omnivores too. These birds also eat the bodies of dead animals they find. Animals that eat dead animals are called scavengers. Scavengers help get rid of dead bodies and **diseases** dead bodies can spread.

Raccoons are nocturnal. That means they're most active at night.

DECOMPOSERS HELP THE DIRT

All food chains end with decomposers. Decomposers are organisms that break down dead, uneaten plants and animals. They get energy as they do this, and they put **nutrients** back into the soil. Plants use their roots to get these nutrients. Without decomposers, plants wouldn't have enough nutrients to live. Without plants, nothing else in a backyard could live!

Most backyard decomposers are bacteria, creatures too small to be seen with human eyes. Worms and ants are larger decomposers that you can see.

Mushrooms, a kind of **fungi**, are also common backyard decomposers. Decomposers are sometimes not found on food chains. They're always the last link, though.

FUN FACT

MOST BACTERIA CAN ONLY BE SEEN THROUGH A MICROSCOPE.

BUILDING A BACKYARD FOOD WEB

Even the smallest outdoor space can support at least part of a food web. You can make a healthy habitat for the animals and plants in your backyard—and the food web there.

Use **native** plants and flowers to attract pollinating bugs. Don't use harmful **chemicals** to kill bugs or weeds. They can hurt other organisms. Make sure there are places for birds to nest and herbivores to find food. The rest of the food web will follow!

Every organism in a food chain or food web matters. If one disappeared, the entire habitat would suffer.

GLOSSARY

algae: Plantlike living things that are mostly found in water.

behavior: The way an animal acts.

carbon dioxide: A gas breathed out by animals.

chemical: Matter that can be mixed with other matter to cause changes.

disease: Illness.

energy: The power or ability to be active.

fungus: A living thing that is somewhat like a plant, but doesn't make its own food, have leaves, or have a green color. Fungi include molds and mushrooms.

habitat: The natural place where an animal or plant lives.

microscope: A tool that makes small objects appear larger when a person looks through it.

native: Having to do with something found naturally in a place, not brought from somewhere else.

nectar: A sweet liquid made by flowering plants.

nutrient: Something taken in by a plant or animal that helps it grow and stay healthy.

reproduce: When an animal creates another creature just like itself.

FOR MORE INFORMATION

Books

Emminizer, Theresa. *Food Chains and Webs.* Buffalo, NY: Enslow Publishing, 2024.

Owen, Ruth. *Food Webs.* Minneapolis, MN: Bearport Publishing Company, 2024.

Websites

Ecosystem: Food Chain and Food Web
www.ducksters.com/science/ecosystems/food_chain_and_web.php
Learn more about these helpful science tools.

Make Your Backyard Wildlife Friendly
kids.nationalgeographic.com/nature/article/make-your-backyard-wildlife-friendly
Check out four easy steps to creating a backyard habitat.

INDEX